50

Things to See with a Small Telescope

John A. Read

www.facebook.com/50ThingstoSeewithaSmallTelescope

Star maps used in this book are made using Stellarium, http://stellarium.org/ an open source stargazing program.

Cover photo by Sean McCauley. Please visit his website (below) for details on how to contact Sean for all your photography and video needs.
http://silhouetteproductions.com

Images of the following telescopes provided compliments of Celestron:
> Celestron First Scope (Page 10), Celestron Powerseeker 114Az (Page 10) and Celestron NexStar 6se (Page 11)

Images of the following telescopes reprinted with permission from Orion Telescopes & Binoculars, www.telescope.com:
> 6 Inch Orion SkyQuest (Page 10), 8 Inch Orion SkyQuest (Page 11)

Image of Meade Lightbridge Dobsonian provided compliments of Meade Instruments

Telescope view source files for deep sky objects were constructed from actual astrophotos used with permission from the following astrophotographers:

> Mark Stanford Sr: Trifid Nebula
> Stuart Forman: Double Cluster, M1, M13, M27, M51, M81 & M82, M81 (Supernova added digitally).
> Mike Harms: Andromeda, Comet, M42

Images from NASA follow NASA's photo usage guidelines found here:
http://www.nasa.gov/audience/formedia/features/MP_Photo_Guidelines.html

Solar and Lunar Eclipse Schedule based on data acquired by Fred Espenak during his time at NASA's Goddard Space Flight Center. Permissions freely granted based the guidelines found here:
http://eclipse.gsfc.nasa.gov/SEpubs/5MCSE.html

*This book is dedicated to Jennifer, who listens to me to talk
about outer space pretty much all the time.*

Acknowledgments

I would like to express my gratitude to Marni Berendsen, developer of the NASA Night Sky Network, for her fantastic contribution editing and fact-checking this book.

I would also like to thank the Mount Diablo Astronomical Society (MDAS) for feeding my desire to learn more about the universe. This book would not be possible without the support of the wonderful folks at MDAS.

To find the astronomy club nearest you, please visit:

https://nightsky.jpl.nasa.gov

Table of Contents

Table of Contents 4

A note from the Author: 6

Introduction 7

Don't have a telescope yet? 10

Difficulty 12

A note on color 13

Things you will need to get going. 14

1. The North Star (Polaris) 15

2. Venus 16

3. Arc to Arcturus then Spike to Spica! 17

4. Betelgeuse 18

5. Rigel 19

6. The Orion Nebula 20

7. Sirius 21

8. The Moon 22

9: Gemini - Castor, Pollux, and Meteors 23

10. Mars 24

11. Jupiter 25

12. Europa 26

13. Io 27

14. Callisto 28

15. Ganymede 29

16. Saturn 30

17. Titan 31

18. Lunar Eclipse 32

18.5. Lunar Eclipse Schedule 33

19. Sunspots 34

20. Solar Eclipse 35

21. The Pleiades 36

22. The Hercules Star Cluster (M13) 37

23. The Milky Way!..38

24. The Andromeda Galaxy 39

25. Comets 40

26. Draco...41

27. Helicopters and Jet Aircraft 42

28. The International Space Station 43

29. Altair and the Summer Triangle...44

30. Cityscapes and Landscapes..45

31. Birds..46

32. The Dumbbell Nebula (M27)..47

33. Albireo 48

34. Mizar & Alcor..49

35. Double Cluster in Perseus 50

36. Vega...51

37. The Ring Nebula 52

38. Meteors, Meteorites, and Meteoroids! 53

39. Asteroids Ceres and Vesta 54

40. The Whirlpool Galaxy (M51) 55

41. Sagittarius Deep Sky Objects...56

42. M81 and M82...57

43. Uranus..58

44. Neptune..59

45. Mercury..60

46. Star-Moon Occultation...61

47. Planet-Moon Occultation 62

48. Iridium Satellite Flares...63

49. The Crab Nebula (M1)..64

50. Supernova 65

Item 51. UFOs..66

Conclusion 67

Appendix 1: Solar Eclipses 2016 – 2021 68

Appendix 2: Solar Eclipses 2022 - 2030......................................69

Appendix 3: Summer Constellation Map for Northern Hemisphere*...........70

Appendix 4: Winter Constellation Map for Northern Hemisphere* 71

A note from the Author:

When I look through my telescope, I am exploring a new and fantastic frontier.

I know you want to skip to the middle of this book, pick something cool, and then try to see it in your telescope. Please note that only about a third of the items in this book will be visible in a given evening. Before you set up your telescope, please download stargazing software for your computer or mobile device. I recommend Stellarium, which is available for free at http://www.stellarium.org or from the app store. Using this software, you can determine if your target is visible. I have chosen difficulty levels for each object (measured in Supernovae). In general, this book is organized in order of increasing difficulty.

Because I do my astronomy in the northern hemisphere, this edition of the book contains several items visible only to those living north of the equator. A Southern Hemisphere edition of this book is also available.

Finally, as the first of many reminders, do not look at the sun through a telescope without utilizing a commercial solar filter. Enjoy!

Introduction

This book is targeted at the small telescope owner. For the purpose of this book, a small telescope is any telescope purchased for a few hundred dollars or less. One of the reasons for writing this book is to address the difficulties met by first time owners of small, department store telescopes.

Many telescopes are used once, packed up, and shoved to the very bottom of a closet. Sometimes, folks are persuaded to purchase these telescopes based on the pictures of planets and galaxies on the box, leading them to believe their new scope is as powerful as the Hubble Space Telescope.

You may have already tried to use the telescope and you realized that the mount is flimsy, the optics are poor, and computer (if it has a computer), which is programed with 14,000 objects, doesn't know Jupiter from the moon.

My first three telescopes met these criteria. As a kid, I spent hours looking at random objects in space, dreaming of someday seeing something exciting. I hoped desperately to see something to ignite my soul, slingshotting me into a lucrative career as an astronaut.

I was an adult before I had one of these enlightening experiences, and well into an established career in corporate accounting when my soul was truly ignited for astronomy. The local pharmacy was selling small telescopes for $13.99. The box was beautifully designed with pictures of Saturn and Jupiter. I thought, *what they heck, I'll do it, I'll buy this telescope!*

I carried the telescope home and set it up. "This scope is really, **not** very good!" I thought, feeling embarrassed for spending money on such a piece of junk. The telescope had a plastic camera tripod instead of a proper telescope mount, the eye pieces were quite small, the primary lens was the size of a large coin and the finder scope was obviously just for decoration.

Anyway, I decided to give it a shot. I carried the telescope outside, setting it up in front of my apartment, under a streetlight and beside the metro-line. I pointed the small telescope at a bright yellow star that had just come up over the horizon.

"Holy Moly!" I thought as the wobbly scope steadied itself in the still air on that clear evening. Before me, in perfect high definition, in perfect focus, without a shimmer of distortion, I saw, for the first time ever, the rings of Saturn.

For many readers, that first telescope you purchased (or received) is a pain in the neck, literally. You have to crane your neck just to look in the eyepiece. Well, this book is for you.

What inspired me to write this book? I do a lot of volunteering with the local astronomical society's outreach group through NASA's Night Sky Network. We go from school to school teaching students about astronomy and how to use a telescope. The thing is, even though we are in California, the sky is not always 100% clear. Here is a common conversation:

Kid: "Can we look at the sun?"

Me: "No, you can only see the sun during the day."

Kid: "Can I see the moon?"

Me: "No, it's not up tonight. But there are plenty of other things to see."

Kid: "Like what?

Meanwhile the clouds begin roll in.

Me: "Like this!" Points telescope at Saturn.

Kid: "I don't see it."

Me: "Ah, a cloud has strategically positioned itself in front of Saturn."

Kid walks away.

When this happens, it's time to get creative, otherwise mayhem follows. The students start to get bored, and they start throwing things. The teachers give them flashlights, which they shine in your eyes. You turn your back for ten seconds and there is the child riding your telescope like a horse.

Sometimes, we just need to think unconventionally. I was on top of Mount Diablo at an astronomy event when the clouds rolled in. I decided to point the telescope at the red light on the top of the observation building at the summit. The students were fascinated!

The light was a quarter mile away, yet you could see the condensation on the red glass enclosure. A moth fluttered around it.

The kids noticed how the light bulb appeared upside down in the scope, and I had to explain how this was due to the lenses and mirrors in the scope. In looking at the light-bulb a quarter mile away, we were able to grasp the power of the telescope, seeing something familiar, something so small, something so far away.

We spent half an hour looking at that light bulb. It was seen by at least a hundred people. That night probably churned out as many future scientists as a night where there were no clouds at all.

Don't have a telescope yet?

Since I published the first version of this book back in 2013, many people have messaged me asking what telescope they should buy given their budget. The most common response to this is "it depends." I hate giving that response. Most people who are getting started in amateur astronomy have one goal: **To see cool stuff**. They're not trying to take pictures, or make groundbreaking scientific discoveries. With this in mind, my one rule for recommending a first telescope is to get the telescope with the most aperture you can afford (aperture is the diameter of the primary lens or mirror).

If your budget is between $25 and $50:

This table top scope has 76mm of aperture, more than enough to see everything in this book. And for around $50, you can't beat the easy to use mount.

Celestron First Scope

Between $50 and $150:

At this price range, start looking for telescopes with over 110mm (~4.5 inches) of aperture. This will enable great views of Saturn's rings, and hundreds of deep sky objects.

Pro tip: Consider getting a used telescope to get more aperture for your buck!

Between $150 and $300:

Celestron PowerSeeker 114AZ

In this range, we're looking at some really great telescopes. Try your best to reach the six inches of aperture range, you won't regret it! Dobsonians make extremely loveable telescopes.

6 Inch SkyQuest

Between $300 and $500:

Now we're talking! Here you can find scopes with between eight and ten inches of aperture. Personally, I prefer Dobsonian for their ease of use, and spectacular views of galaxies, nebulas and globular clusters.

8 Inch Orion SkyQuest

Between $500 and $1000

At this price range, you may want to consider trading aperture for a computerized telescope. I personally wouldn't, but it is an option. A twelve inch Dobsonian is a serious telescope. In dark skies, you can see distant comets, and dim galaxies. Some people even use these telescopes to search for undiscovered supernovae!

Meade Lightbridge Dobsonian

Under $1000, the go-to, or computerized telescopes tend to have no more than six inches of aperture. However, many go-to scopes have cool features like tours of the sky and satellite tracking.

Celestron NexStar 6se

11

Difficulty

Here is a helpful guide to the level of difficulty required to view each object.

1 Supernova

Seriously, how have you not seen this before?

2 Supernovae

Probably one of the brightest objects in the sky

3 Supernovae

If you can see this, you're officially an amateur astronomer!

4 Supernovae

Real astronomers envy your accomplishment*

5 Supernovae

You've just discovered a real supernova and are suddenly a media darling!

*Sometimes it can take hours of patience to finally find the object you are looking for and it may not always be spectacular, but that's not the point. The point is to appreciate the objects that you can see! Hopefully this book will help you appreciate the true splendor of everything in the sky.

A note on color

Did you know that in dim lighting the human eye can only see in black and white?

Only when you use a digital camera do galaxies and nebulas get color. Many objects imaged using professional telescopes aren't even in wavelengths the human eye can see! In this case, professional astronomers assign a color the human eye *can* see to that particular wavelength of light. This is called false color, or representative color.

This book is about what **you** can see through your telescope, not what a camera can image. Astronomers who focus on visual astronomy often refer to "beautiful smudges," because without a camera, that is what most deep sky objects look like.

For this reason, this book is different than most other beginner astronomy books. I've chosen to keep the print version in Black and White, which saves you, the budding astronomer, almost $15 that you can now put towards your new telescope!

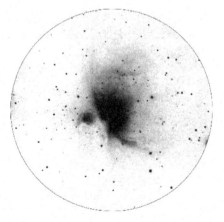

A Beautiful Smudge!

Things you will need to get going.

1. That telescope you got for Christmas (or Hanukkah, or your birthday).

2. A basic understanding of how to focus and point it at the bright stuff in the sky. See you telescope's manual for more details.

3. A stargazing application such as "Stellarium" for Mac and PC available at http://www.stellarium.org or at the app store. You will need a stargazing application to determine the location for many of the objects mentioned in this book. For the most part, planets do not follow any annual calendar, so you will need software to find a planet's current location in the sky.

4. You will need a commercial solar filter if you plan on using your telescope to look at the sun. When looking at the sun ALWAYS use a commercial solar filter over the **objective lens** or **primary mirror**. These filters can be purchased at an online telescope retailer such as http://www.telescopes.com.

Never use a solar filter that covers only your eyepiece. The sunlight will burn through the filter and YOU WILL IMMEDIATELY GO BLIND.

1. The North Star (Polaris)

Many people have incorrect assumptions about which star is actually the North Star. Some people believe that it's the brightest star in the sky. I've actually had folks argue with me over which star is the North Star, some people even pointing to Sirius (located on the other side of the sky) just because it was the brightest star they could see at the time. In reality, the North Star is the 48th brightest star in the night sky.

To find the North Star, follow the two stars that form the front of the cup of the Big Dipper to the next brightest star (as shown in the diagram below).

The North Star is what is commonly called a visible binary star. With your telescope you may be able to make out the second star, Polaris B.

Polaris is very important to folks who own an equatorially mounted telescope in the Northern Hemisphere. In order for this type of mount to function correctly, one axis must be pointed directly at this star.

Difficulty: 1 Supernova

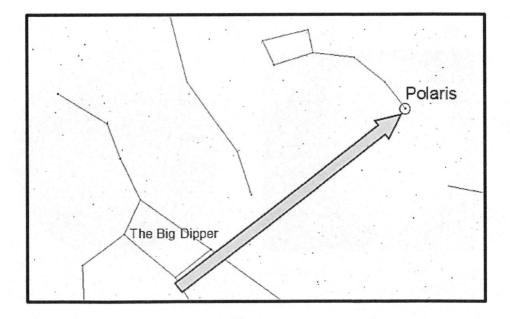

2. Venus

Ah, Venus! This beautiful planet is named after the Roman Goddess of Love and Beauty. Venus is closer to the Sun than the Earth. For this reason, you can only see Venus shortly after sunset or right before sunrise.

Venus is bright. So bright in fact, that Venus is one of the primary sources of UFO sightings amongst pilots. This is due to an optical illusion. Large objects viewed at great distance don't appear to move along with an observer (the person who is viewing the object). This creates the illusion that the observer is being followed by the object; in this case, Venus.

As mentioned above, Venus can either be seen just before sunrise, or just after sunset. To find Venus, use the program Stellarium to determine its specific location.

Through a telescope Venus looks a bit like our moon. It appears white and even has phases. This is because Venus is closer to the Sun and we sometimes see Venus's nighttime side.

When someone else looks though your telescope and says, "Hey, I see the moon!" just ask them to step back and have a look at where the telescope is pointed.

Difficulty: 1 Supernovae

Venus imaged by the Mariner 10 Spacecraft

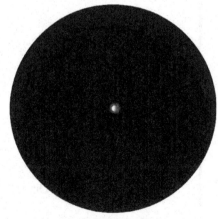

Venus through a telescope

3. Arc to Arcturus then Spike to Spica!

Beginning in the springtime, "Arc to Arcturus then Spike to Spica" is a great phrase to remember as you begin to navigate around the eastern sky. By creating an arc with the handle of the big dipper and following it across the sky to arrive at the bright star Arcturus, you can then straighten your arc to hop over to the bluish star, Spica.

Arcturus is an Orange Giant and the fourth brightest star in the sky. Spica is a Blue Giant and the fifteenth brightest star. Spica resides in the constellation Virgo, while Arcturus is located in Boötes, (which is much more fun to say).

Arcturus is very interesting because over the course of our lifetime it will move relative to nearby stars (about one seventh the diameter of the moon in one hundred years). The star is moving at over ninety miles per second, so fast that in 500,000 years, it will be gone from sight altogether!

Spica is both rotating and variable (increases and decreases in brightness). At its equator, it rotates at almost two hundred kilometers per hour and changes in brightness ever so slightly with each rotation.

Difficulty: 1 Supernova

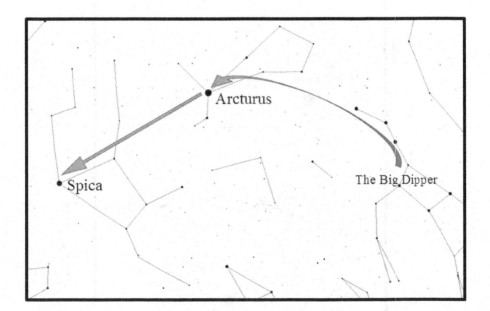

4. Betelgeuse

Yes Betelgeuse, somewhere in the vicinity of which, *The Hitchhiker's Guide to the Galaxy* is said to have been written! Kids love this star, mainly because it sounds like Beetlejuice (a film inspired by the star's name).

This big red star surprises those who think all stars are white (including me until a few years ago). Betelgeuse also varies in brightness over time. It is often the 8th brightest star in the sky, but it can be as bright as the 6th, or as dim as the 20th!

Betelgeuse can be found near the top of the Orion constellation. When looking at it through a telescope it's easy to see how red it is. To contrast its redness, pan the telescope down to Rigel, a blue star detailed in the next section.

Objects in the Orion constellation are best viewed in the fall and winter months. Most people find Orion by locating the three bright stars that make up Orion's belt.

Difficulty: 1 Supernova

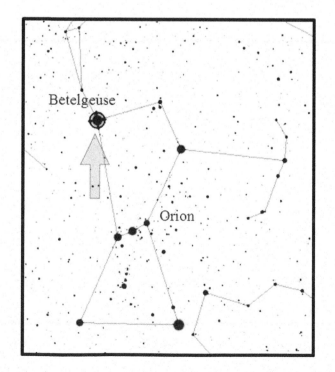

5. Rigel

Not one, not two, but three stars make up this point of light found on Orion's foot. If you have very dark skies it's possible to separate star A (a Blue Supergiant) and star B (a much dimmer companion star). However, star C orbits very close to star B, and is impossible to separate using a small telescope.

If Rigel is really three stars, it must have several planets, right? The writers of Star Trek seem to think so. Planets named Rigel X, Rigel II or Rigel VII make the Rigel system about the most popular place in the Star Trek Universe!

As of March 2016, no planets have been discovered around Rigel. However, thousands of new planets are being found each year. You can find an updated database of these discoveries at http://exoplanets.org/.

While observing, remember to contrast Rigel's color and brightness against Betelgeuse.

Difficulty: 1 Supernova

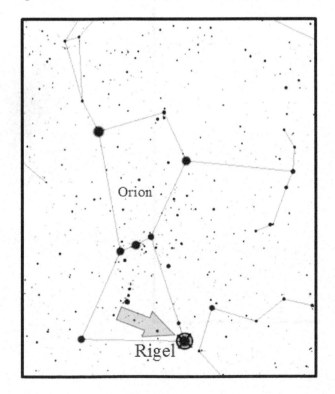

6. The Orion Nebula

The Orion Nebula is often dubbed the "Star Factory." When you observe this nebula you can see a great expanse of gas surrounding a series of stars. It is called the "Star Factory" because these stars are being formed out of that gas.

The Orion Nebula is part of the Orion Molecular Cloud Complex, which also contains the Horsehead Nebula. Although the Horsehead is far too dim to see in a small telescope, it is nonetheless the location of "The Planet of the Ood" from BBC's classic series *Doctor Who.*

The Orion Nebula is one of the easiest to find deep-sky objects (objects not located in our Solar System). Find Orion's belt, and then picture his sword as the line of stars running down from the belt. The middle of this sword contains the Orion Nebula.

Difficulty: 2 Supernovae. Finding the Orion Nebula is like riding a bike. You never forget how to do it.

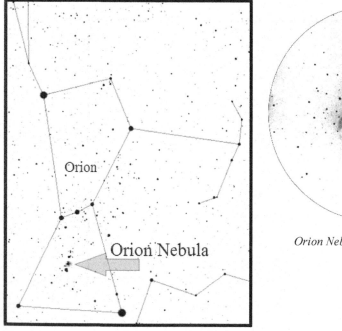

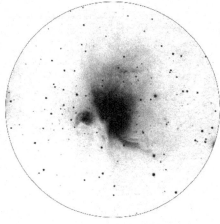

Orion Nebula through a telescope

٦. Sirius

Sirius is the first stop of the Harry Potter tour (many star names and constellations are mentioned in the Harry Potter books). This star is twice as bright as any other star in the sky and will effectively ruin your night vision for up to thirty minutes! It is so incredibly bright that at high altitudes it can be seen during the day.

This star is nicknamed the "Dog Star" due to its prominence in the constellation Canis Major (Greater Dog) and inspired the phrase, "Dog days of summer." In Harry Potter, the character Sirius Black transforms into a dog. Coincidence? I don't think so.

Sirius is located to the left of the Orion constellation and can be seen prominently in the southern sky during the fall, winter and early spring.

Difficulty: 1 Supernova

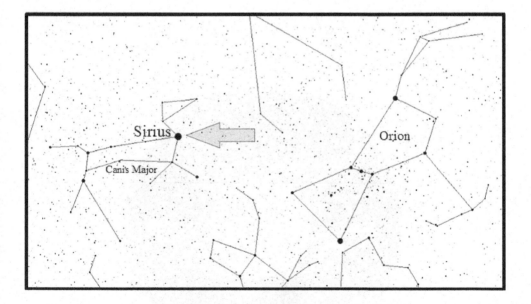

8. The Moon

You can't miss it! With even the smallest scopes, you should be able to clearly see the craters on the surface.

I once used that telescope purchased at the pharmacy for $13.99 to try to film NASA's "Lcross" mission. During this mission NASA crashed a spacecraft into the moon. Scientists were attempting to create a plume of moon dust they could then analyze for traces of water. The crash was supposed to create a flash of light visible from Earth, but I didn't see anything. However, the crash was not visible because the spacecraft (which crashed into a southern crater) impacted into lunar soil with the consistency of snow!

The moon is visible for about half the month in the evening sky. If you really think about it, this makes sense because, as most of us know, the moon orbits the Earth every 27 days. I am often surprised when, on moonless nights, some folks seem to think that we can see the moon with the use of a telescope. Just to clarify, if you can't see the moon without a telescope, you can't see it with one.

Difficulty: 1 Supernova

The moon through a small telescope

9: Gemini - Castor, Pollux, and Meteors

The constellation Gemini is best seen in the winter and spring in the western sky after sunset and is visualized by picturing twins holding hands. Stars Castor and Pollux make up the heads of these twins.

The star Castor, the head of the rightmost twin, is a double star when viewed through the telescope. But Castor is actually a sextuple star system, six stars bound together by gravity. These six stars can only be separated by an extremely strong telescope, or through the science of spectroscopy (breaking down light into different wavelengths).

The star Pollux, the head of the leftmost twin, used to be a "main sequence star" like our sun. However, it burned through its hydrogen and has since expanded into a "giant" star many times the radius of our sun. This gives the star its orange color. Pollux is also the brightest known visible star with a planet orbiting it (this may change as new planets are discovered).

The Geminid meteor shower in mid-december is one of the most prolific meteor showers of the year. Be sure to give your eyes plenty of time to adapt to the dark, this way you'll see even more shooting stars.

Difficulty: 2 supernovae

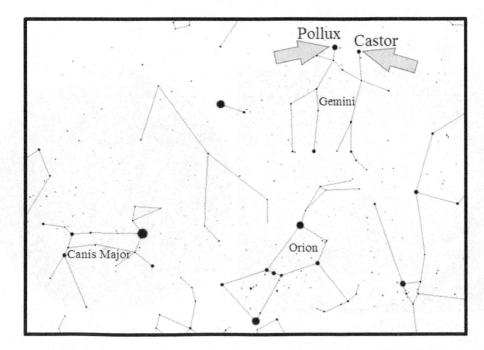

23

10. Mars

Sure it might look like just a simple red disk in your telescope, but hey, it's Mars! Keep looking and keep focusing and you may be able to see the polar ice caps and some varying colors in the Martian soil.

It is very cool to realize that there are men and women here on Earth (At NASA's Jet Propulsion Laboratory in Los Angeles County) remotely piloting rovers the size of small SUVs and golf carts on the surface of Mars.

Since Mars is a planet, it will be found along the ecliptic*. As with all planets, check astronomy software like Stellarium for a precise location. If you already know Mars is visible, check the ecliptic for a deep red looking star.

*What is the ecliptic? Since all the planets travel around the sun in approximately the same orbital plane, they will all appear in a specific slice of the night sky; sort of like an airplane that always takes the same route. This path is called the ecliptic and it roughly runs from the eastern horizon to the western horizon. This is also the path the sun follows during the day.

Difficulty: 2 Supernovae

Mars imaged by Hubble

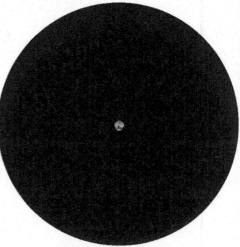

Mars through a telescope

11. Jupiter

If you want to be impressed, take a look at Jupiter and its four largest moons: Europa, Io, Ganymede and Callisto! For half the year, Jupiter is one of the very first things to show up in the night sky after dark, this makes it a great target for focusing your telescope and aligning your finder scope.

Jupiter is a huge planet, over two and a half times the mass of all the other planets in the solar system combined. With a small telescope in good focus, not only should you be able to see the four moons discovered by Galileo in 1610, you may also see the two most pronounced cloud belts on the planet itself.

To find Jupiter, look for one of the brightest objects in the sky on the ecliptic (the path of the planets through the sky from east to west), or simply check Stellarium or other astronomy software. Use a medium powered eyepiece for optimal viewing.

As you can see from the children's photos below, Jupiter is also a great target for practicing astrophotography!

Difficulty: 2 Supernovae

Planet Jupiter Photographed by Children aged 3 -12
Lafayette Library Astroblast - Jan 30th, 2013

12. Europa

The moons of Jupiter need their own section because they are just so interesting.

Europa is the smallest of the four moons discovered by Galileo. It also has liquid water; lots of liquid water. Latest estimates project that beneath an icy surface, there is an ocean over 60 miles deep. By this estimate, Europa has twice as much water as there is on Earth!

Jupiter's moons change position every night. For the most part, it's difficult to tell which moon is which using a small telescope. The best way to tell which moon is Europa is by using astronomy software.

Difficulty: 3 Supernovae

Jupiter and its moons - (moon orientation changes every night)

Europa imaged by the Galileo Spacecraft

13. Io

Have you read the book *Ilium* by Dan Simmons? Well, you should because one of the main characters (a mining robot) is from this moon.

Of Jupiter's moons discovered by Galileo, Io is the one that orbits most closely to Jupiter. Io is also the most geologically active body in the solar system sporting over 400 active volcanos!

Due to the amount of volcanic activity, Io's surface features frequently change. Using large telescopes like the Keck Telescope in Hawaii, professional astronomers can monitor the volcanic activity and the changes the volcanoes make to the surface. Whereas most moons in the solar system are covered with meteorite impact craters, Io has almost none. This is because lava flows cover them up soon after they are formed.

Difficulty: 3 Supernovae

Io imaged by the Galileo Spacecraft

14. Callisto

Pack your bags, because Callisto could be your new home! This moon has the lowest radiation levels of Jupiter's large moons, and would make a promising location for human settlement! That is, if you can stand days that are 400 hours long. If you ever visit Callisto, don't try to stay up all night!

When looking at Jupiter, Callisto is usually the moon that appears farthest from the planet. It orbits so far that it's easy to confuse it with a background star.

Difficulty: 3 Supernovae

Callisto

Callisto imaged by the Galileo Spacecraft

15. Ganymede

Made famous by the 1993 television series "Power Rangers" this moon hosted the location of the Zord fleet of Mega Vehicles. How would you like to get that Jeopardy question, eh?

More interestingly, Ganymede is the largest moon in the solar system having over twice the mass of Earth's moon!

To find Ganymede, look closely to see which of Jupiter's moons is the biggest and brightest. But, to be sure, check your astronomy software for confirmation.

Difficulty: 3 Supernovae

Ganymede imaged by the Galileo Spacecraft

16. Saturn

One look at Saturn and you might trade in your car for a telescope of equal value. Or not. Either way, it's quite a sight.

In fact, Saturn is so awesome, the most awesome day of the week is named after it. That's right, Saturday, or as you should call it from now on, Saturn-is-Awesome-Day.

As with any planet, first check Stellarium or another stargazing app to make sure that it is high in the night sky. It will be along the ecliptic and will appear yellow in color.

Difficulty: 2 Supernovae (3 Supernovae if you can take a photo of the rings with your camera-phone).

Saturn imaged by the Cassini Spacecraft

Saturn through a telescope

17. Titan

Titan is the largest moon of Saturn. What better place to drop out of warp to avoid detection from a Romulan mining vessel in the fantastic movie *Star Trek 11*.

Most interesting about Titan is that the gravity is low enough and the atmosphere thick enough, that by attaching small wings to your arms, you could fly like a bird!

On January 14th, 2005 NASA landed a small probe named *Huygens* on Titan. *Huygens* penetrated Titan's thick atmosphere and parachuted to the ground. The probe took photos all the way down and one photo from the surface (shown to the right).

Because Titan has a thick atmosphere, it also has an interesting climate. The surface temperature is almost -180 °C and it often rains liquid methane. In addition to rain, radar images of the planet's surface confirm the existence of hydrocarbon seas and lakes.

To find Titan, first, find Saturn. Once you have found Saturn, Titan will be orbiting right beside it.

Difficulty: 3 Supernovae

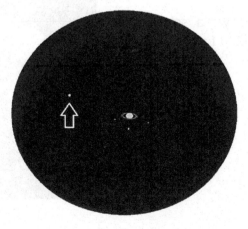

Saturn and Titan through a telescope

18. Lunar Eclipse

Often referred to as a Blood Moon, Lunar eclipses are not as rare as you might think. Unlike solar eclipses which are only visible in certain places, Lunar Eclipses can be observed from almost anywhere on the nighttime side of Earth, assuming there are no clouds in the way.

A lunar eclipse occurs when the moon passes into the shadow of the Earth. Sunlight passes through the Earth's atmosphere giving the moon a reddish hue.

There are three basic types of Lunar Eclipses. First, and most exciting, is the Total Lunar Eclipse, where the moon is totally immersed the Earth's Shadow. Second, there is the Partial Lunar Eclipse. During a Partial Eclipse the moon is only partially covered. Finally, there is the Penumbral Lunar Eclipse, where light passing through the Earth's atmosphere illuminates a section of the moon, but no distinct shadow is visible. However, Penumbral Eclipses are often difficult to distinguish from a regular full moon.

The next page shows a schedule of Total and Partial Lunar Eclipses through the year 2030.

Difficulty: 2 Supernovae

Lunar Eclipse, Author's Photo

18.5. Lunar Eclipse Schedule

Calendar Date	Eclipse Type	Greatest Eclipse Time (UT ~ UTC)	Eclipse Duration	Geographic Region of Eclipse Visibility
August 7, 2017	Partial	18:21:38	01h55m	Europe, Africa, Asia, Australia.
January 31, 2018	Total	13:31:00	03h23m	Asia, Australia., Pacific, Western North America
July 27, 2018	Total	20:22:54	03h55m	South America, Europe, Africa, Asia, Australia.
January 21, 2019	Total	5:13:27	03h17m	Central Pacific, Americas, Europe, Africa
July 16, 2019	Partial	21:31:55	02h58m	South America, Europe, Africa, Asia, Australia.
May 26, 2021	Total	11:19:53	03h07m	East Asia, Australia, Pacific, Americas
November 19, 2021	Partial	9:04:06	03h28m	Americas, Northern Europe, East Asia, Australia, Pacific
May 16, 2022	Total	4:12:42	03h27m	Americas, Europe, Africa
November 8, 2022	Total	11:00:22	03h40m	Asia, Australia, Pacific, Americas
October 28, 2023	Partial	20:15:18	01h17m	Eastern Americas, Europe, Africa, Asia, Australia
September 18, 2024	Partial	2:45:25	01h03m	Americas, Europe, Africa
March 14, 2025	Total	6:59:56	03h38m	Pacific, Americas, Western Europe, Western Africa
September 7, 2025	Total	18:12:58	03h29m	Europe, Africa, Asia, Australia
March 3, 2026	Total	11:34:52	03h27m	East Asia, Australia, Pacific, Americas
August 28, 2026	Partial	4:14:04	03h18m	East Pacific, Americas, Europe, Africa
January 12, 2028	Partial	4:14:13	00h56m	Americas, Europe, Africa
July 6, 2028	Partial	18:20:57	02h21m	Europe, Africa, Asia, Australia
December 31, 2028	Total	16:53:15	03h29m	Europe, Africa, Asia, Australia, Pacific
January 26, 2029	Total	3:23:22	03h40m	Americas, Europe, Africa, Middle East
December 20, 2029	Total	22:43:12	03h33m	Americas, Europe, Africa, Asia
June 15, 2030	Partial	18:34:34	02h24m	Europe, Africa, Asia, Australia

Eclipse Predictions by Fred Espenak, NASA's GSFC

19. Sunspots

Sunspots are eddies or storms of magnetic activity near the surface of the sun. These storms reduce the surface temperature over a given area and create the dark spots visible in solar filtered telescopes.

What's cool about sunspots? Well, first, they are usually about the size of the Earth! Second, they come in pairs (one for each magnetic pole of the disturbance). Third, they change locations every day. Fourth, I once took a photo of a sunspot that looked like Hawaii.

To view sunspots, use a commercial solar filter over your telescope or binoculars, and then get the sun in good focus. With the sun in focus, you should almost always be able to see at least one or two sunspots.

Difficulty: 2 Supernovae

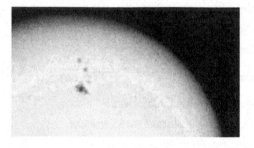

Sunspots that look like Hawaiian Islands

Photographing the sun using solar filtered binoculars and an iPhone

20. Solar Eclipse

A Solar Eclipse occurs when the moon passes in front of the Sun. Due to the elliptical orbit of the moon, sometimes the eclipse happens when the moon is closer to the Earth, and sometimes it happens with the moon is farther away. For this reason, there are two types of eclipses. First, there is the Annular Eclipse, where the moon is farther away and cannot completely cover the sun. Second, there is the Total Solar Eclipse, when the moon orbits close to the Earth and fully blocks out the sun.

I admit that I will not witness a Total Solar Eclipse until the next one in 2017, but I hear that viewing a Total Solar Eclipse is an amazing experience; the air gets cooler, animals do strange things, and it gets considerably darker.

I have only experienced an annular eclipse, which is how I was able to take the photo below (using my iPhone, binoculars, and a solar filter).

For the hour before and hour after totality you can view the sun through your telescope using a commercial solar filter. Totality is when the moon fully covers the sun. This can last anywhere from thirty seconds to six minutes.

A schedule for all Solar Eclipses through the year 2030 is included in the appendix of this book.

Difficulty: 2 Supernovae

Annular Solar Eclipse - May 20th, 2012

21. The Pleiades

You can skip this one if you drive a Subaru, because you see this star cluster every time you look at your steering wheel. If you don't drive a Subaru, then the Pleiades can be found to the right of Orion, (that's your right, Orion's left). According to Greek mythology the Pleiades or Seven sisters were turned into stars by Zeus to help them flee Orion, who, ironically, still pursues them through the night sky.

Some people think that this cluster of stars is the Little Dipper. It's not. The actual Little Dipper is quite dim, yet considerably larger than the Pleiades and is located in the northern sky.

To find the Pleiades, look up and to the right of Orion, usually, with any amount of light pollution, only 6 of the brightest stars in the Pleiades are visible to the unaided eye. However, as soon as you look in your telescope, dozens of stars will appear!

Difficulty: 1 Supernova

The Pleiades through a telescope

22. The Hercules Star Cluster (M13)

This Globular Cluster is one of only a few objects in this book that resides outside the plane of the Galaxy! It is also where the Earth was hidden in Dan Simmon's classic (1989) novel *Hyperion* (sorry for the spoiler).

M13 is one of the brightest deep sky objects. It's also very easy to find, because this star cluster is huge, containing several hundred thousand stars. If you are using binoculars or a very small telescope, M13 will appear as a grey glob (hence globular).

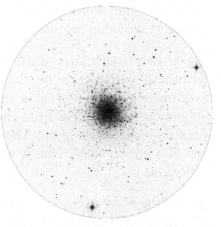

To find M13, pan around the edge of the square in the constellation Hercules until you find it.

Difficulty: 3 supernovae.

Hercules Star Cluster through a telescope

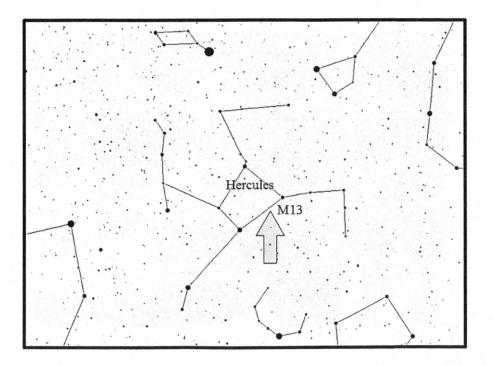

23. The Milky Way!

If you're an amateur astronomer (if you own a telescope, that's you) and you can't find the Milky Way, well, I guess you just need darker skies! In fact, all of the stars that you see in the night sky are part of the Milky Way. When someone says they see the Milky Way, they are actually referring to the *plane* of the Milky Way. You can clearly see the plane in the photo included below.

If you live near a large town or city, you probably cannot see the white wispiness that makes up the plane of the Milky Way. In fact, the maximum number of visible stars in the sky from within a large city is only about a dozen. Far from city lights, you might count as many as 6000 stars on a moonless night. The Milky Way contains between 300 billion and 400 billion stars! That is why it appears as a white wispiness in truly dark skies.

One of the ways to explore the plane of the Milky Way with a telescope is to start at one horizon and work your way across to the other. You never know what you'll find.

Difficulty: 1 Supernova

Milky Way from Hawaii. Author's photo.

24. The Andromeda Galaxy

Before the twentieth century, the Milky Way was thought to be the only galaxy in the universe. Astronomers dubbed objects that seemed to reside outside of the galaxy "Island Universes." It wasn't until Edwin Hubble measured the distance to the Andromeda Galaxy that debate over the Island Universes closed. Before Hubble, many astronomers believed that the Andromeda Galaxy was actually a nebula and called it *The Andromeda Nebula*.

The cool thing about the Andromeda Galaxy is that it is over six times as wide as the full moon! However, the only way to see the full extent of this galaxy is through long exposure photography. When you see the Andromeda Galaxy in your telescope, you only see the bright galactic core, which appears as a beautiful grey smudge.

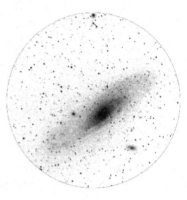

Andromeda Galaxy through a telescope

To find the Andromeda Galaxy, use the constellation Cassiopeia and observe the distance between any two stars in the W, then count over three of these lengths as shown in the diagram below.

Difficulty: 3 Supernovae

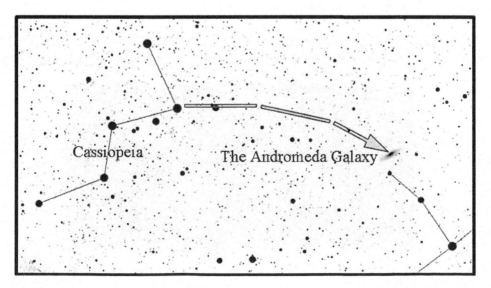

25. Comets

What's the best way to find out if you can see a comet? Read the news. Approaching comets usually get picked up by the media. However, in the media, vastly exaggerated claims of brightness (or apocalyptic close encounters with Earth) are common. Despite the hype, only a few of these comets can actually be glimpsed by the casual sky watcher.

Comets are city-sized balls of ice often traveling over one hundred thousand kilometer per hour. When passing in the vicinity of the Sun, comets "out-gas" creating a visible tail of particles millions of miles long.

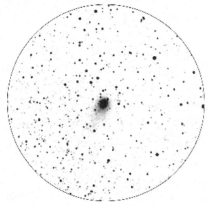

We observe comets from a distance of hundreds of millions of miles. Though they are traveling at great speed, they are often visible for as long as a month. This gives the amateur astronomer plenty of time to observe.

Comet through a telescope

How to see a comet: Astronomy websites and even the media will post prominent stories when a comet is visible in the night sky. Most of these sources will provide instructions on where to look. If the comet is dim, use binoculars to scan the sky according to the map, once you have found it, move to your telescope for a closer look.

Difficulty: 2-5 Supernovae depending on the comet, 2 if the comet can be seen with the unaided eye, and 5 if you discover a new comet and get to name it!

Comet with the unaided eye

26. Draco

Yes, this is another stop along the Harry Potter astronomical tour. But since all of the stars in the constellation Draco are pretty dim, they are not the reason that this item is on this list.

If you know Latin, then you know Draco means dragon. If you look at the constellation, you will see the dragon's head and every October, this dragon breathes fire! The October Draconids are meteors that appear to shoot from the head of the dragon.

For a cool photo, put your camera on a tripod and take 30 second exposures all night long. If you don't have a camera with manual exposure, use the fireworks setting. You might just get a newsworthy photo of this real fire-breathing dragon.

Difficulty: 1 Supernova for finding the constellation, 4 supernovae for photographing a meteor.

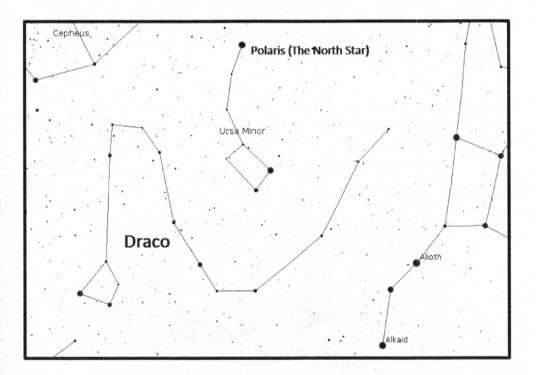

27. Helicopters and Jet Aircraft

Do you live in a high crime area? I sure do. The next time the police are searching for the culprit, use your telescope to see if you can differentiate the police chopper from the news chopper.

You might think that this item is strange to be included in an astronomy book, however, the world's greatest astrophotographers such as Thierry Legault use aircraft as practice in preparation for spotting fast moving objects in space such as the International Space Station. Thierry's amazing work can be found here: http://legault.perso.sfr.fr/.

To see a plane in your telescope, you'll want to use the minimum amount of magnification; this will require the use of your largest eyepiece. Use the finder scope to narrow in on the plane and begin to move your scope to keep it in view. Keep tracking as you move from the finder scope to the eyepiece.

Tracking an aircraft is easier or harder depending on the type of mount you are using. A Lazy-Susan mount (called a Dobsonian), alt-azimuth mount, or camera mount will be optimal; whereas an equatorial mount will be difficult as movement is restricted.

Chasing jet aircraft is a great star party activity for children before it gets dark. Just make sure the sun has set so you don't accidently point the scope in that direction. When I work with students, we sometimes play a game to see who can guess which airline the plane belongs to. Then, we look in the telescope to find out!

Difficulty: 2 Supernovae

Space Shuttle Endeavour and Carrier Aircraft. Photo by Author.

28. The International Space Station

Dubbed "ISS" by those in the space community, the International Space Station can be seen at least a few times per week from almost every location on Earth. It is visible either in the morning before sunrise or in the evening shortly after sunset.

Viewing the Space Station with your telescope can be tough, especially if you have an equatorial mount, but with a Dobsonian or tabletop design, it can be a relatively easy target. Search the internet or use the NASA app to determine when the International Space Station will pass over your location.

To see the ISS in your scope, use an eyepiece that provides medium magnification. Track the station in your finder and then switch to the eyepiece. If you are lucky, you should be able to make out the solar panels.

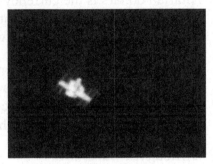

ISS. Author's Photo

How is it possible to see so much detail? The ISS is orbiting only a few hundred miles above the Earth and is the size of a football field. This means the Station can appear three times as large as Saturn!

Note: Acquiring the ISS in your scope is much easier with two people, one to track the space station in the finder scope and one person to observe the station through the eyepiece.

Difficulty: 4 Supernovae

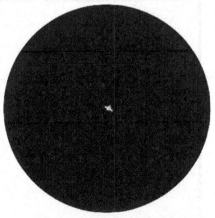

(Note: the ISS moves VERY fast)

29. Altair and the Summer Triangle

The Summer Triangle (or as my wife calls it, "The Great Pizza Slice") is an interesting part of the sky because it straddles the plane of our galaxy. Because of this, it is filled with many objects to discover as you dive deeper into astronomy and upgrade to larger telescopes.

The Summer Triangle is a great way to learn your way around the sky. It is outlined by three stars: Vega, Deneb, and Altair.

Altair is frequently mentioned in science fiction due to its proximity to Earth. At only 16.7 light years away, it is one of the closest bright stars. Altarian dollars is the currency used throughout *The Hitchhiker's Guide to the Galaxy*. Altair is also mentioned in multiple Star Trek episodes, *Star Trek*, *The Wrath of Khan and* two episodes of *Doctor Who*.

No planets have been discovered orbiting Altair, but this may change in 2017 with the launch of a spacecraft called TESS (Transiting Exoplanet Survey Satellite). TESS will continually scan two million of the closest star systems searching for Earth-like planets.

Difficulty: 1 Supernova

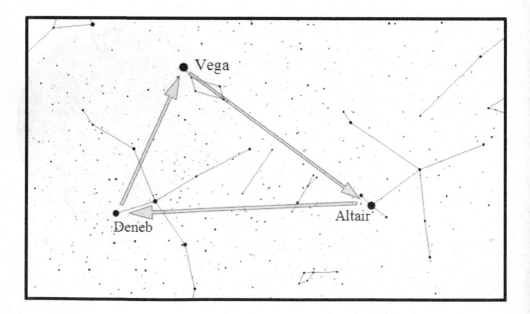

30. Cityscapes and Landscapes

Pointing the telescope at ground-based objects is a great way to learn the power of your telescope. I was volunteering at an event on Mount Diablo in California when we pointed the telescope toward San Francisco. The Giants had just won their game and fireworks were going off above the stadium. You couldn't see this without the telescope, so all the kids that night gathered around took turns watching the fireworks.

The challenge with looking at ground-based objects is that most telescopes invert the image. For this reason, some telescopes use an "inverting" lens to turn things right side up.

Landscapes become great telescope targets. This is why many tourist attractions have permanently mounted telescopes or binoculars at every lookout.

If you are at Yosemite, check out the climbers scaling El Capitan. If you are camping at the Lava Beds National Monument, check out miles and miles of volcanic rock. Camping on the beach? Use your telescope to observe the ships out at sea.

You might even see a whale!

Difficulty: 1 Supernova

Golden Gate Bridge from Mount Diablo. Author's Photo.

31. Birds

Personally, I don't know much about birds, but some folks purchase their telescopes with bird watching in mind. Some small telescopes, like the Meade ETX 60, come with a separate camera slot for this purpose.

One of the great things about viewing birds with a telescope is the depth of field. Depth of field is a term used in photography to describe the degree to which the subject is in focus. When viewing a bird on a tree with a telescope, only the bird will be in focus. This is because the telescope naturally creates a "shallow" depth of field.

Telescopes are best for viewing birds that are far away. For birds that are close, it is better to use binoculars. According to the internet, the best birds to look at through a telescope are wildfowl in the open country, or seabirds.

Difficulty: 2 Supernovae, if there are lots of birds. 4 supernovae if there are very few birds.

Bird in Berkeley. Author's Photo

32. The Dumbbell Nebula (M27)

Discovered in the year 1764 by the French astronomer Charles Messier, the Dumbbell Nebula was the first planetary nebula ever discovered. It also has a very large apparent size in the telescope. The photo below shows its apparent size relative the moon.

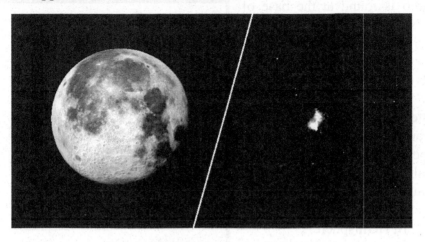

Moon and M27 at the same magnification

M27 is located in the **Summer Triangle** between the constellations Vulpecula and Sagitta.

Interestingly, the Dumbbell Nebula wasn't given its name until 1833 when astronomer John Herschel made this record:

"A nebula shaped like a dumb-bell, with the elliptic outline completed by a feeble nebulous light."

Difficulty: 3 Supernovae

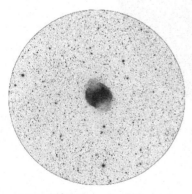

Dumbbell through a Telescope

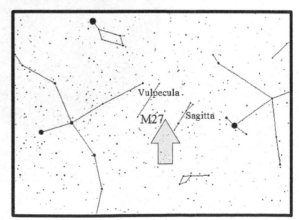

33. Albireo

Albireo is a star-party favorite. This is because you can see a vast contrast between two colors of stars. Albireo is a binary star in which Albireo A is a yellow star and Albireo B is a blue companion.

Albireo is found at the base of the Northern Cross, which is actually not a constellation but an asterism (an asterism is an easily recognizable group of stars, but not officially a constellation. Another example of an asterism is the Big Dipper). This constellation is actually Cygnus, The Swan. Cygnus is primarily a summer and fall constellation in the Northern Hemisphere.

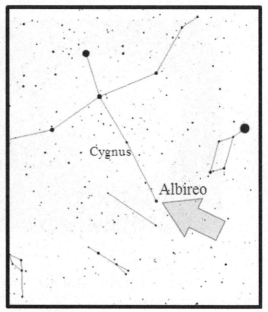

Difficulty: 2 Supernovae

Albireo through a telescope (In this image, the yellow star is on the left)

34. Mizar & Alcor

No need for optometrists when you have these two stars in sight. Nicknamed the "Horse and Rider" seeing these stars located in the Big Dipper used to be a test of eyesight! These days, most people can make out these two stars with corrected lenses. Mizar and Alcor make up the center of the handle of the Big Dipper.

In addition to being a visual double star, Mizar was the first telescopic double star every observed. It was discovered by an Italian mathematician named Benedetto Castelli in 1617.

When observing these stars, first notice the double stars which can be seen with the unaided eye, then look at the stars through the telescope. Mizar will be the star with the companion so close they appear to almost touch.

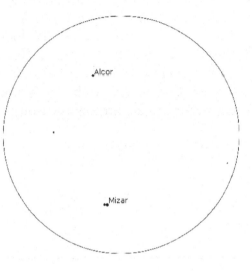

Mizar and Alcor through a telescope

Difficulty: 2 Supernovae

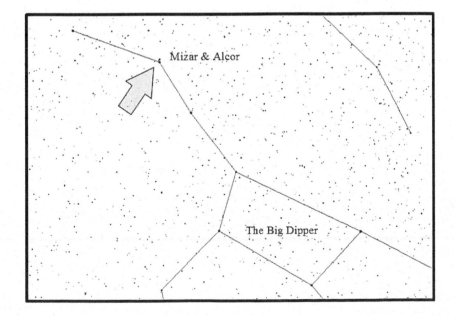

35. Double Cluster in Perseus

These star clusters are notable for two reasons. First, they are easy to find from the Northern Hemisphere since they are above the horizon most evenings of the year. Second, each year the Perseid Meteor showers originate from this part of the sky in mid-August.

Star clusters are great for showing just how many stars are out there! To find the double Cluster in Perseus, look to Cassiopeia (the big W) and find the clusters below and to the left of the W (or up and to the right of a big M depending on the time and season).

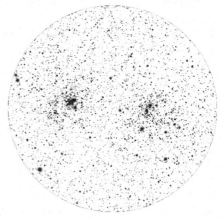

Difficulty: 2 Supernovae.

Double Cluster through a telescope

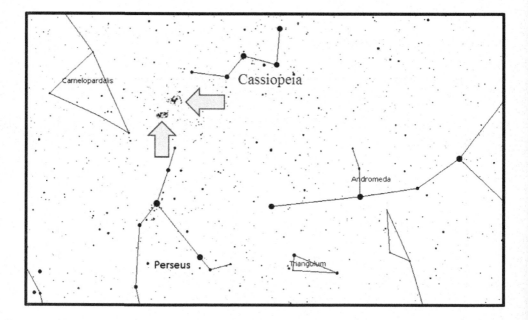

36. Vega

Yes, Jodie Foster's home planet; just kidding (The extraterrestrial radio array from the book and movie *Contact* is located at Vega).

Interestingly, Vega was the North Star about twelve thousand years ago and it will be again about twelve thousand years from now. This is due to the Earth's precession around its axis.

Precession is a property of rotating objects. You can observe precession directly in spinning toys such as a gyroscope or a top. A gyroscope will precess, if you tap it, by way of a smooth wobble. For the Earth, precession is mainly the result of the gravitational influence of the Sun and the Moon.

Vega is the brightest star in the constellation Lyra, and is visible high in the sky during the summer. Also in this constellation is the famous Ring Nebula (as discussed in the next section).

Difficulty: 1 Supernova

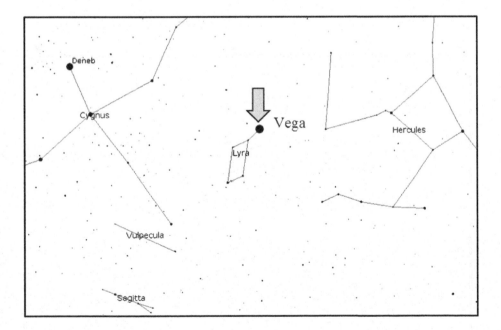

37. The Ring Nebula

The Ring Nebula appears about as large as Jupiter in your telescope, but not nearly as bright. The challenge in a small telescope is to clearly make out the hole in the Ring. In order to view the center of the Ring, you will need a telescope with a lens or mirror of at least 10cm (4 inches) in diameter.

This Nebula was formed when a Red Giant star shed its outer shell of ionized gas, leaving only a white dwarf star where the Red Giant once was.

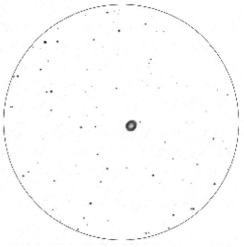

To find the Ring Nebula, pan the telescope between the stars Sheliak and Sulafat in the constellation Lyra.

Difficulty: 3 Supernovae

Ring Nebula through a telescope

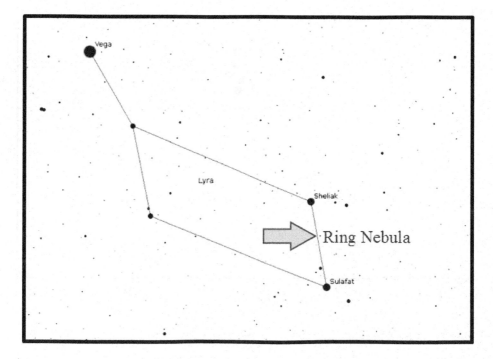

38. Meteors, Meteorites, and Meteoroids!

Meteors, Meteorites, and Meteoroids! Even I get these terms confused. A meteoroid is the term for the small rock while it is in space. Once the rock enters the atmosphere, it becomes a meteor or "shooting star." A good way to remember this is that we have "meteor showers", not meteorite showers. The space rock is only called a meteorite if it touches the ground. You'll probably never see a meteoroid in the telescope because of their small size. Usually, if the rock is bigger than a few meters across, it would be classified as an asteroid.

If you do any amount of stargazing, you will see plenty of meteors, I guarantee it. Just this past Friday I was working with a school group in Walnut Creek, California, when a very bright meteor passed by the section of sky we were all watching. You could see the meteor break up and fizzle over the period of a few seconds.

Most meteors are smaller than a golf ball. You can see them because they are moving at tens of kilometers per second and when they hit the atmosphere, they burn very brightly.

You will even see meteors in your telescope! You can't plan on seeing a meteor, but with enough time looking, you are bound to see one cross your field of view.

Difficulty: 1 Supernova without a telescope, 3 supernovae if you are lucky enough to have a meteor cross your field of view while looking through your telescope.

Author holding a meteorite

39. Asteroids Ceres and Vesta

You may know about the asteroid belt between Mars and Jupiter, but most folks don't realize the sparseness of the belt. Even in the asteroid belt, space is very, very empty. The asteroid Ceres makes up a third of all the mass in the entire asteroid belt. The mass of all the asteroids combined is less than 4% of the mass of our Moon!

In 2006 the International Astronomical Union reclassified Ceres as a Dwarf Planet (like Pluto). Vesta, due to its smaller mass, is classified as a minor-planet.

To see Ceres or Vesta, use astronomy software the same as you would a planet. Ceres and Vesta are quite small and look like stars through a telescope, so if you're not sure which point of light is the asteroid, sketch the location of the brightest stars in that area. Observe the same location in a few days and the asteroid is the object that moved. Ceres and Vesta can even be seen without a telescope in extremely dark skies.

Vesta Imaged by the Dawn Spacecraft

Difficulty: 4 Supernovae

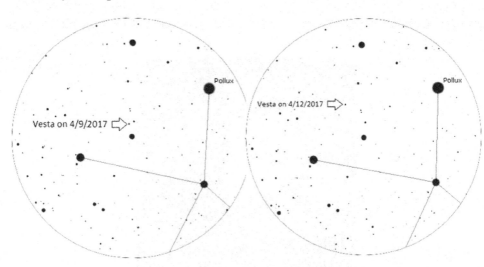

Vesta moving from one night to the next

40. The Whirlpool Galaxy (M51)

The Whirlpool Galaxy, or M51, is easy to find in a small telescope or even binoculars, but only on moonless nights and far from city lights. This Galaxy is accompanied by a smaller companion galaxy designated NCG 5191 or M51b. The gravitational interaction between these two structures is thought to give the Whirlpool its well defined spiral shape.

Astronomers have discovered that most large galaxies have a supermassive black hole at their center and observations of M51 by the Hubble telescope reveal a distinct X shaped pattern around this galaxy's center. One bar of the X is most likely dust circling the black hole. The second bar of the X could be dust interacting with a cone of ionized particles. Further observation is required before astronomers reach a scientific consensus.

Supernovae have also been observed in this galaxy in 1994, 2005 and 2011.

To find the Whirlpool Galaxy, make a right triangle under the handle of the big dipper as shown below.

Difficulty: 4 Supernovae

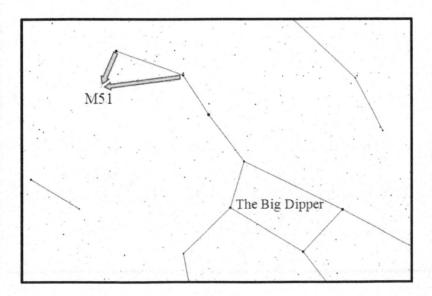

Whirlpool Galaxy through a telescope

M51

The Big Dipper

41. Sagittarius Deep Sky Objects

Even as an amateur astronomer, I'm not used to picturing the entire Sagittarius constellation. Fortunately, there is an asterism (an unofficial constellation) called the Teapot, which I consider to be Sagittarius (see image).

Sagittarius is a great place to explore deep sky objects (objects outside of our Solar System) because it lies in the direction of the center of our Milky Way Galaxy. This constellation is a fun place to explore because there is a good chance of finding something interesting, even without using a star map. For example, you might find the Lagoon Nebula, the Omega Nebula or the Trifid Nebula.

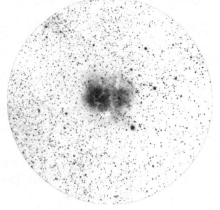

To see all the good things in Sagittarius, use an eyepiece without much magnification since most of the items here are fairly large. Scan the upper right of the Teapot to find nebula, and scan around the remainder of the Teapot for star clusters.

Difficulty: 3 Supernovae

Trifid Nebula through a telescope

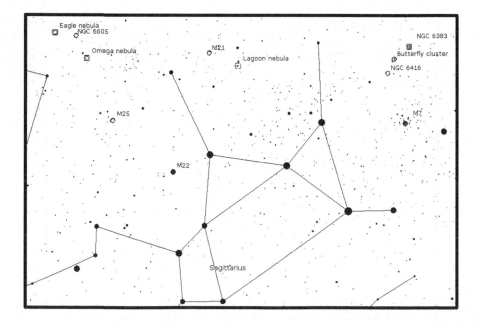

42. M81 and M82

After Andromeda, M81 and M82 are the two galaxies that are the easiest to find. M82 is commonly referred to as the Cigar Galaxy because of how it appears from Earth. M81 is also referred to as Bodes Galaxy.

M81 is particularly interesting to professional astronomers because in its center is a gigantic black hole with a mass 70 million times that of our sun!

To view these galaxies, use an eyepiece with low magnification. With the Big Dipper as a guide, create a line between the bottom left of the Dipper's cup and its lip. Then extend this line from the lip to arrive at the location of these galaxies.

M81 and M82 through a telescope

Difficulty: 4 Supernovae

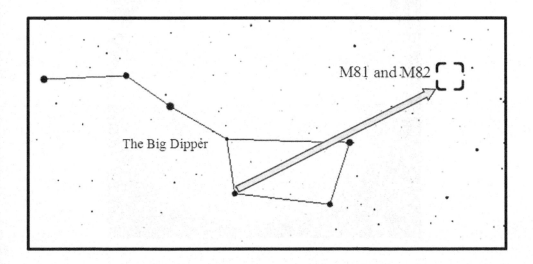

43. Uranus

To English speakers, the most interesting thing about Uranus is its name, which shares the pronunciation with a certain private piece of human anatomy. Though it sounds funny to some people, its name is very logical. Saturn is the father of Jupiter and Uranus is the father of Saturn. It's one big family out there in the outer solar system.

Because Uranus is so far from the sun, it will stay in relatively the same part of the sky for our entire lives. For the twenty-first century, the best time to view Uranus is in September and October.

To find Uranus, first check your astronomy software to find the precise location. Use a low magnification eyepiece to make the initial find, and then move to a higher magnification eyepiece to resolve the planet and more of the planet's hue.

Difficulty: 4 Supernovae

Uranus imaged by the Voyager 2 Spacecraft

44. Neptune

Now that Pluto has been demoted to a "Dwarf Planet" by the Astronomical Union, Neptune is the farthest planet from the Sun (in our solar system). Like the other planets (besides Earth) this one is named after a Roman god, the God of the Sea.

Neptune is very dim, one of the dimmest objects in this book. However since it is blue, it can be distinguished from background stars. As with Uranus, use an eyepiece without much magnification to find the planet. Then, use an eyepiece with high magnification to get a better view. Note, only telescopes that are six inches in diameter or larger will be able to resolve Neptune into a disk. For smaller telescopes, the planet will appear as a point of light.

Difficulty: 4 Supernovae

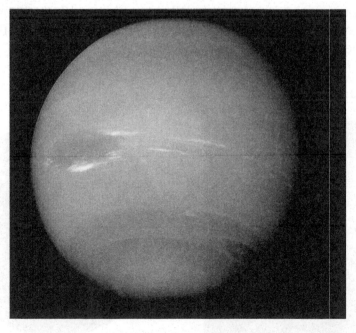

Neptune imaged by the Voyager 2 Spacecraft

45. Mercury

Due to Mercury's extreme closeness to the Sun, this planet can be extremely challenging to get a good look at. It may only appear in the evening sky a few days per year. As with Venus, you see Mercury in phases. These phases greatly affect its brightness. When Mercury is visible, it is only visible for a very short time before sunrise and after sunset.

To find the best time to see Mercury, use astronomy software such as Stellarium, click and lock (hit spacebar) onto Mercury. Then, use the "time" setting to fast forward until Mercury is above the horizon after sunset. You can also pay attention to astronomy websites, they'll often let you know when Mercury is visible.

When observing Mercury through your telescope, it may look extremely bright and even shimmering as if it is on fire. Mercury's apparent brightness is due to its proximity to the sun, but the shimmering is due its apparent proximity to the horizon. When you view objects that are low in the sky, you are looking through more atmosphere than when the objects are overhead. The atmospheric distortion makes the object appear to shimmer.

Difficulty: 4 Supernovae

Mercury imaged by the Messenger Spacecraft

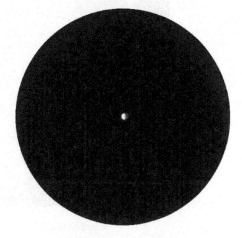

Mercury through a telescope

46. Star-Moon Occultation

Occultations occur when one object goes behind another in space. Sort of like an eclipse. The most common occultations are when the moon passes in front of a bright star.

Grazing occultations tend to be the most interesting. This is when a star appears to graze the surface of the moon from your location. During a grazing occultation, it's not uncommon for the star to blink in and out of sight as it goes between mountain ranges or gullies on the surface of the moon.

This is a great chance to use the "time" feature of your astronomy software." To find out when an occultation will occur (without consulting astronomical journals, magazines, or websites) just open your astronomy software and select the moon.

After the moon has been selected, it should lock to the center of your screen (try hitting the spacebar if you are using "Stellarium"). Then, using the time feature, begin to run the "hours" into the future. You should see the stars moving in the background while the moon stays in place. You may have to fast forward for a few weeks before the moon occults a bright star. When it does, mark your calendar and set a reminder to watch a star disappear behind the moon.

Difficulty: 4 Supernovae

47. Planet-Moon Occultation

As mentioned in the previous section, an occultation occurs whenever two objects align so that one is covering the other from the perspective of the observer. For example, if Saturn passes behind the moon, you would say, "Saturn has been occulted by the moon" (almost sounds like it should be a crime).

To find a planetary occultation, use the same technique as used for star occultations. With the moon selected in your software, run the hours forward for a few days, weeks or months, until you see the moon pass directly over a planet. Then, set a reminder, and wait for the event to occur.

Taking a photo of this with a smart-phone is difficult, but not impossible. To take a photo with your smart-phone, place the camera up to the eyepiece, then tap on the image of the moon. This should set the focus and exposure. Then, take the photo! If you get a good photo, immediately post it on www.spaceweather.com. By posting it here, your photo may end up on CNN or other major news networks!

Difficulty: 4 Supernovae

Jupiter-Moon Occultation

48. Iridium Satellite Flares

A normal satellite in orbit, when viewed from Earth, is about as bright as a dim star. Satellites can be observed moving quickly across the sky shortly after sunset, or before sunrise. However, if that satellite is an Iridium Communications Satellite with shiny antennas, then you might be in for a treat!

The easiest way to spot the flares from Iridium Communication satellites is by downloading a phone app such as Sputnik: http://sputnikapp.info The app creates a forecast for your location and sends you alerts when there is about to be a flare.

You do not need a telescope to view these flares, but it may be fun to use a telescope anyway. Viewing moving objects in space is good practice for when you want to view something more challenging like a near Earth asteroid or the International Space Station.

Difficulty: 3 Supernovae

Iridium Flare over San Francisco. Author's Photo.

49. The Crab Nebula (M1)

Something special happened on the fourth of July, 1054 and it wasn't an American Independence Day celebration. On this day, Chinese astronomers recorded what they thought was a new star, a star brighter than Venus! After a few weeks however, the new star dimmed, but it was still visible for almost two years, at which point the star was nearly lost to history.

The story could have ended there, but in 1731, almost seven hundred years later, a British Astronomer named John Bevis observed a blob in that exact spot. Then, almost three decades after that, a French comet hunter named Charles Messier added this "blob" to his catalog of objects that were "Definitely not comets." Messier designated the object "M1." In other words, the blob was the number one item on his list of "Non-comets."

The Crab Nebula is a Supernova remnant. The Chinese observed the actual supernova, the violent explosion of a star. Now, when you look through your telescope, you are observing the ongoing explosion of dust and gasses shooting through space at almost five million kilometers per hour.

To find the Crab Nebula, search the area right above Orion's head.

Difficulty: 3 Supernovae

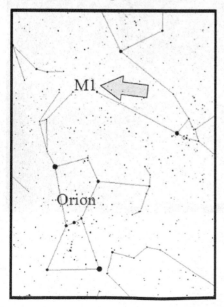

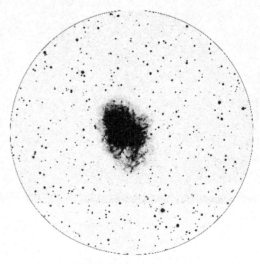

Crab Nebula through a telescope

50. Supernova

If you are looking at a galaxy and realize that it has a new "star" in it, you may have spotted a supernova! Supernovae are caused when a star explodes and releases enough energy to outshine an entire galaxy.

When a star goes supernova, particles called neutrinos are released in the hours before the explosion. These neutrinos are detected by instruments around the Earth, giving an approximate location of the supernova to come. A message goes out via the internet to members of the astronomical community and the hunt is on! If you are the only person to observe the supernova, your name makes the news.

If, however, the supernova has already been discovered, you can get its location from the internet and attempt to see it for yourself!

Difficulty: 5 Supernovae

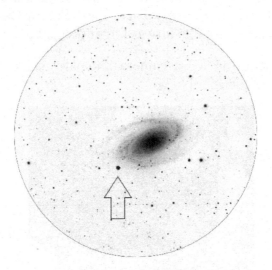

Supernova through a telescope

Item 51. UFOs

Every year there are tens of thousands of reported UFO sightings. These are usually recorded by folks who are unaccustomed to observing the sky, or who review their film or camera footage and see something they don't understand.

UFO sightings are usually explained by common optical illusions, or phenomenon within the camera equipment. But it's still exciting to observe something you don't understand. Many folks in the United States live near military bases and regularly see strange things in the sky (probably drones).

I saw my first "UFO" when I was delivering papers as a young boy. I was standing beside a farmer's field at 5a.m. when a bright light rose from behind a distant hill. I stopped and watched the bright light grow in size until it was almost blinding. For another five minutes the light persisted, moving back and forth in the sky. Then, the UFO, (a Dash 8 Series 100 Aircraft) flew overhead with its forward light pointed in a new direction.

Difficulty: 0 Supernovae for a camera anomaly and 6 Supernovae for getting abducted by aliens.

Lens flair within the Camera

Conclusion

I hope you enjoyed this tour of *50 Things to See with a Small Telescope!* If you would like to continue in this hobby, I would highly encourage you to join your local astronomical society. A list of these clubs within the United States can be found here:

http://nightsky.jpl.nasa.gov/club-map.cfm

If you love fiction, check out my science fiction thriller, The Martian Conspiracy.

"A hard sci-fi novel with shades of Kim Stanley Robinson's *Red Mars,* though much faster-paced. If, like me, you dream of living on Mars you should read this book."

-Graeme Shimmin, Author of: *A Kill in the Morning*

67

Appendix 1: Solar Eclipses 2016 – 2021

Type	Date	Time of Greatest Eclipse (UTC)	Location
Total	March 9, 2016	1:58:19	**Total:** Indonesia, Micronesia,Marshall Islands **Partial:** Southeastern Asia, Korea, Japan, Eastern Russia, Alaska, Northwestern Australia, Hawaii,Pacific
Annular	September 1, 2016	9:08:02	**Annular:** Atlantic, Central Africa,Madagascar, Indian Partial: Africa, Indian Ocean
Annular	February 26, 2017	14:54:33	**Annular:** southern Chile and Argentina, Angola, southwestern Katanga Partial: Southern and Western Africa, Southern South America,Antarctica
Total	August 21, 2017	18:26:40	Total: Oregon, Idaho, Wyoming,Nebraska, Northeastern Kansas,Missouri, southern Illinois, Western Kentucky, Tennessee, Southwestern North Carolina, Northeastern Georgia, South Carolina **Partial:** North America, Hawaii,Greenland, Iceland, British Isles, Portugal, Central America,Caribbean, northern South America, Chukchi Peninsula
Partial	February 15, 2018	20:52:33	Partial: Antarctica, Southern South America
Partial	July 13, 2018	3:02:16	**Partial:** South Australia, Victoria,Tasmania, Indian Ocean, Budd Coast
Partial	August 11, 2018	9:47:28	**Partial:** Northeastern Canada,Greenland, Iceland, Arctic Ocean,Scandinavia, northern British Isles, Russia, Northern Asia
Partial	January 6, 2019	1:42:38	**Partial:** Northeastern Asia, Southwestern Alaska, Aleutian Islands
Total	July 2, 2019	19:24:08	**Total:** central Argentina andChile, Tuamotu Archipelago **Partial:** South America, Easter Island, Galapagos Islands, Southern Central America,Polynesia
Annular	December 26, 2019	5:18:53	**Annular:** northeastern Saudi Arabia, Bahrain, Qatar, United Arab Emirates, Oman,Lakshadweep, Southern India, Sri Lanka, Northern Sumatra, southern Malaysia, Singapore,Borneo, central Indonesia, Palau,Micronesia, Guam **Partial:** Asia, Western Melanesia, Northwestern Australia, Middle East, East Africa
Annular	June 21, 2020	6:41:15	**Annular:** Democratic Republic of the Congo, Sudan, Ethiopia,Eritrea, Yemen, Empty Quarter,Oman, southern Pakistan, Northern India, New Delhi, Tibet, southern China, Chongqing,Taiwan **Partial:** Asia, Southeastern Europe, Africa, Middle East, WestMelanesia, Western Australia,Northern Territory, Cape York Peninsula
Total	December 14, 2020	16:14:39	**Total:** Southern Chile andArgentina, Kiribati, Polynesia **Partial:** Central and SouthernSouth America, Southwest Africa,Antarctic Peninsula, Ellsworth Land, Western Queen Maud Land
Annular	June 10, 2021	10:43:07	**Annular:** Northern Canada,Greenland, Russia **Partial:** Northern North America,Europe, Asia
Total	December 4, 2021	7:34:38	**Total:** Antarctica **Partial:** South Africa, South Atlantic
Eclipse Predictions by Fred Espenak, NASA's GSFC			

Appendix 2: Solar Eclipses 2022 - 2030

Type	Date	Time of Greatest Eclipse (UTC)	Location
Partial	April 30, 2022	20:42:36	**Partial:** Southeast Pacific, Southern South America
Partial	October 25, 2022	11:01:20	**Partial:** Europe, Northeast Africa,Mid East, West Asia
Hybrid	April 20, 2023	4:17:56	**Hybrid:** Indonesia, Australia,Papua New Guinea **Partial:** Southeast Asia, East Indies, Philippines, New Zealand
Annular	October 14, 2023	18:00:41	**Annular:** Western United States,Central America, Colombia, Brazil **Partial:** North America, Central America, South America
Total	April 8, 2024	18:18:29	**Total:** Mexico, Central United States, East Canada **Partial:** North America, Central America
Annular	October 2, 2024	18:46:13	**Annular:** Southern Chile, Southern Argentina **Partial:** Pacific, Southern South America
Partial	March 29, 2025	10:48:36	**Partial:** Northwest Africa, Europe, Northern Russia
Partial	September 21, 2025	19:43:04	**Partial:** South Pacific, New Zealand, Antarctica
Annular	February 17, 2026	12:13:06	Annular: Antarctica **Partial:** South Argentina, Chile,South Africa, Antarctica
Total	August 12, 2026	17:47:06	**Total:** Arctic, Greenland, Iceland,Spain, Northeastern Portugal **Partial:** Northern North America,Western Africa, Europe
Annular	February 6, 2027	16:00:48	**Annular:** Chile, Argentina, Atlantic Partial: South America,Antarctica, West and South Africa
Total	August 2, 2027	10:07:50	**Total:** Morocco, Spain, Algeria,Libya, Egypt, Saudi Arabia,Yemen, Somalia **Partial:** Africa, Europe, Mid East, West and South Asia
Annular	January 26, 2028	15:08:59	**Annular:** Ecuador, Peru, Brazil,Suriname, Spain, Portugal **Partial:** Eastern North America, Central and South America, Western Europe, Northwest Africa
Total	July 22, 2028	2:56:40	**Total:** Australia, New Zealand Partial: Southeast Asia, East Indies
Partial	January 14, 2029	17:13:48	**Partial:** North America, Central America
Partial	June 12, 2029	4:06:13	**Partial:** Arctic, Scandinavia,Alaska, Northern Asia, Northern Canada
Partial	July 11, 2029	15:37:19	**Partial:** Southern Chile, SouthernArgentina
Partial	December 5, 2029	15:03:58	**Partial:** Southern Argentina, Southern Chile, Antarctica
Annular	June 1, 2030	6:29:13	**Annular:** Algeria, Tunisia,Greece, Turkey, Russia, Northern China, Japan **Partial:** Europe, Northern Africa,Mid East, Asia, Arctic, Alaska
Total	November 25, 2030	6:51:37	**Total:** Botswana, South Africa,Australia **Partial:** South Africa, SouthernIndian Ocean, East Indies,Australia, Antarctica
Eclipse Predictions by Fred Espenak, NASA's GSFC			

Appendix 3: Summer Constellation Map for Northern Hemisphere*

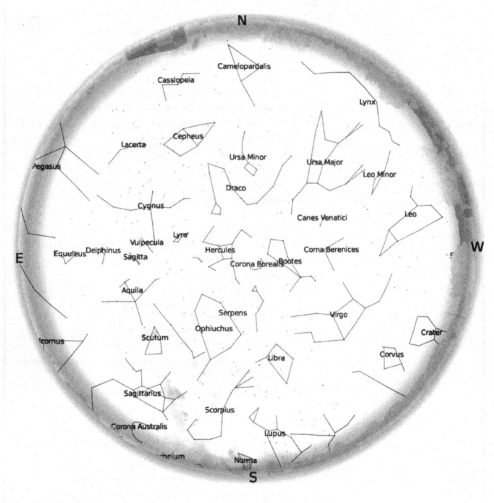

*Latitude 37 degrees

Appendix 4: Winter Constellation Map for Northern Hemisphere*

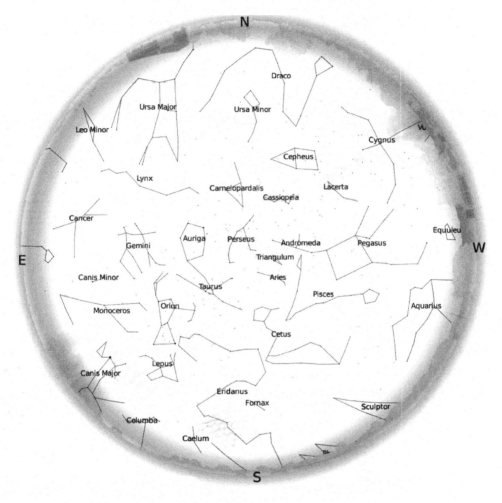

*Latitude 37 degrees

CPSIA information can be obtained
at www.ICGtesting.com
Printed in the USA
BVOW04s0509201216
471293BV00057B/913/P